FOOD COLORS

Committee on Food Protection
Food and Nutrition Board
Division of Biology and Agriculture
National Research Council

NATIONAL ACADEMY OF SCIENCES
Washington, D.C.
1971

NOTICE: The study reported herein was undertaken under the aegis of the National Research Council with the approval of its Governing Board. Such approval indicated that the Board considered the problem of national significance and that the resources of NRC were particularly suitable to the conduct of the project.

The members of the committee were selected for their individual scholarly competence and judgment with due consideration for the balance and breadth of disciplines. Responsibility for all aspects of this report rests with the study committee, to whom sincere appreciation is expressed.

Although the reports of committees are not submitted for approval to the Academy membership or to the Council, each is reviewed according to procedures established and monitored by the Academy's Report Review Committee. The report is distributed only after satisfactory completion of this review process.

ISBN 0-309-01930-3

Available from

Printing and Publishing Office
National Academy of Sciences
2101 Constitution Avenue, N.W.
Washington, D.C. 20418

Library of Congress Catalog Card Number 70-176233

Printed in the United States of America

Preface

The consumer is interested in, and increasingly aware of, the use of chemicals and other aids in processing the many foods commercially available to him. The purpose of this monograph is to provide a general perspective on the use of one class of these food chemicals, i.e., coloring agents, for the consumer and for professionals who work with consumers.

Committee on Food Protection

William J. Darby, Chairman
David B. Hand, Vice Chairman
John C. Ayres
Julius M. Coon
Kenneth P. DuBois
Lloyd J. Filer, Jr.
Leon Golberg
Wayland J. Hayes
Lloyd W. Hazleton
Emil M. Mrak
Bernard L. Oser
Virgil B. Robinson
John H. Rust
R. Blackwell Smith, Jr.
Frank M. Strong

Subcommittee on Food Technology

David B. Hand, Chairman
Emil M. Mrak, Vice Chairman
John C. Ayres
Douglas G. Chapman
Richard L. Hall
John R. Matchett
Reid T. Milner
Willard B. Robinson
Arthur T. Schramm
Harold W. Schultz
Kenneth G. Weckel

Contents

Contents

1
Introduction

The appearance of a food is important in determining its acceptability. Generally, before a food is ingested, "we taste it with our eyes" and accept or reject it almost at a glance. We have a mental index of the red, orange, yellow, green, blue, or violet pigments that we tend to associate with meats, citrus, other fruit and fruit juices, berries, butter, cheese, olives, pickles, syrups, baked goods, or confections.

Certain foods have been artificially colored for many years, and we have come to expect them to have a characteristic color intensity and hue. Those that do not are often considered unacceptable, however wholesome they may be. And we have come to assume that such foods as beverages and fruit drinks, gelatin deserts, maraschino cherries, candied fruits, jellies, candies and other confections, ice cream, sherbets, breakfast cereals, bakery goods, snack foods, and pet foods will be appropriately colored. A wide array of coloring agents is needed for this purpose including—in addition to reds, oranges, yellows, greens, blues, and purples—those that impart browns, whites, grays, and blacks. A single color does not necessarily produce the same effect in all foods, and desired hues and intensities are often achieved only by blending two or more colors. Dependability from one time to the next and uniform dispersion of the color within the food are important.

Foods may be exposed to air, light, and heat during processing, packaging, and storage, and it is important that the color of the product remain uniform and stable under these conditions. In every instance, the material must be safe. Small wonder that the task of producing the desired colors for the food industry is formidable.

Synthetic food colors are used in most countries. Approximately 65 such colors are known to be in use, but the number permitted in the various countries varies widely: 33 in Denmark; 25 in the United Kingdom; 22 in Japan; 22 in the European Economic Community countries, 12 in the United States (not including several synthetic colors that are not subject to certification); 10 in Canada; 5 in Chile; 3 in the Soviet Union; and none in Greece. Discussion in this monograph is limited to those that may be used in the United States.

Average annual per capital consumption of food in the United States is about 1,420 lb (645 kg). Of this amount, synthetic food colors constitute only about 0.012 lb (5.5 g), nearly 85 percent of which is made up of amaranth (FD&C Red No. 2), tartrazine (FD&C Yellow No. 5), and sunset yellow FCF (FD&C Yellow No. 6).

Colors that are used only in very small amounts are difficult to maintain on the market and are frequently produced by color manufacturers as a service to the processor rather than as a profit item. The delisting of a color for reasons of safety necessitates a search for an acceptable replacement, involving costly research and testing. Nonetheless, it is obvious that colors must be examined carefully and withheld from the food supply if they impose a hazard.

The use of colors in foods probably antedates the written record. We know, from Pliny, that wine was colored as far back as 200–300 B.C., and as early as 1396 an edict was issued in Paris that forbade the coloring of butter. There are few records on the composition of early food colors, but they must have been either mineral pigments or substances of animal and vegetable origin.

Beginning in the early nineteenth century, as the food processing industry developed, the artificial coloring of foods received increasing attention, and accounts from that period are both interesting and frightening. For example:

- In 1820 Frederick Accum mentioned the fate of a woman who customarily ate pickles while at her hairdressers—pickles that were colored green with copper sulfate—and who became ill and died.
- Cheese colored with vermillion (HgS) and red lead (Pb_3O_4) also quite understandably caused illness.
- In a Manchester "tea" shop, stocks of copper arsenite, lead chromate, and indigo were found on hand to color used tea leaves for resale.
- Candies were generally colored; in the absence of synthetic dyes, mineral pigments were commonly used. Hassall analyzed 100 samples of candy and found that 59 contained lead chromate; 12, red lead; 6, vermillion; and 4, white lead (basic lead carbonate).
- In Boston in 1880, 46 percent of the candy sampled contained one or more mineral pigments, chiefly lead chromate.

- In 1860, a caterer wishing to have a green pudding at a public dinner asked a druggist to provide a color, and the copper arsenite he received and used caused two deaths.
- In London about 1900, the addition of yellow coloring to milk was so common that housewives refused to buy uncolored milk, thinking it had been adulterated. The yellow tint was commonly added to prevent detection of skimmed or watered milk. It was not until 1925 that a British law prohibited the coloring of milk.

It is understandable that attitudes today toward legitimate use of color in foods sometimes reflect the fact that for over 2000 years a common purpose of adding color was to defraud the consumer or to disguise adulteration.

A number of food colors used in the United States, i.e., pigments that occur naturally in plants and animals or that are simple organic or inorganic compounds, are exempt from the certification procedure established for most of the synthetic materials.

2
History of Food Color Certification Prior to the Food, Drug, and Cosmetic Act of 1938

Synthetic dyes began to attract attention in the food industry soon after Sir William Henry Perkin synthesized mauve in 1856. These synthetic dyes were far superior to many of the vegetable and mineral dyes in tinctorial value, stability, and uniformity. In addition, they were available in many additional hues. In the United States, Congress authorized the addition of color to butter in 1886, and by 1900 synthetic dyes were used in a great many food products. In 1900, the then Bureau of Chemistry of the U.S. Department of Agriculture was given funds to investigate the nature of food preservatives and coloring matters and their relationship to health. This led to the classic work of Dr. Bernhard C. Hesse (1912).

Hesse was a German dye expert specifically employed for this project. He proposed that any synthetic dye could be used for food, provided it was found to be harmless and necessary. At that time there were 695 different coal-tar dyes on the market and little was known about most of them.

In an effort to limit his task, Hesse first set out to discover the need for specific dyes. He wrote to the 37 manufacturers of these dyes requesting each to furnish information on and samples of any dyes regarded as suitable for food and asking that each sample be identified by number according to the 1904 edition of A. G. Green's *A Systematic Survey of the Organic Coloring Matters*. Seventy percent of the manufacturers responded, providing 284 samples, 172 of which were identified by Green table numbers and 6 by unequivocal chemical technology. Only these 178 samples were further considered by Hesse, who found that they represented only 80 different chemical compounds.

Hesse then studied available published data on the physiological behavior of the 80 compounds, taking into account the opinions of qualified experts and the laws of such other countries as Austria, Germany, Italy, France, and Belgium. Contradictions cropped up in the literature and even in legal codes of different countries. After tabulating all recorded observations, he retained for study only those dyes for which no unfavorable report had been made. Sixteen dyes on the United States market (in 1907) met this criterion. Upon these, physiological data were obtained, for the most part, by testing for acute, short-range effects in dogs, rabbits, and human beings.

The next limitation set by Hesse was concerned with the shades required and the general demand as reflected by requests from various sources. Of the 16 dyes, 10 were red shades representing three classes of chemical structures: azo, triphenylmethane, and xanthene. Amaranth, requested by seven sources, was selected from the azo group and erythrosine, requested by five, from the xanthenes. One dye each from the yellow, green, orange, and blue shade groups was similarly selected. Hesse then added one more, Ponceau 3R, on the basis of requests from four sources for red monoazo compounds.

Seven dyes finally selected by Hesse were recognized under the Pure Food and Drugs Act of 1906 in the first Food Inspection Decision on dyes issued on July 13, 1907. These are given in Table 1. Blends of these dyes were used to produce intermediate shades.

Hesse summarized his work as follows: (a) Coal-tar dyes should not be used indiscriminately in foods; (b) only specified coal-tar dyes should be used in foods; and (c) only tested and certified dyes should be used in foods. In selecting other dyes for food use, Hesse proposed rules that required physiological testing in man and animals before acceptance. He also emphasized the need for strict control in the manufacture of these dyes to ensure a pure, clean product.

On the basis of Hesse's recommended rules for selection, 10 dyes were added to the approved list between 1916 and 1929 (see Table 2).

Under the Pure Food and Drugs Act of 1906, provision was made for the certification of all food dyes by the Secretary of Agriculture, but on a voluntary basis only. The Federal Food, Drug, and Cosmetic Act of 1938, however, made certification mandatory for the 15 colors then on the list. The recent history of food colors is involved deeply in legislative and regulatory matters at both the national and international levels; these involvements are discussed in greater detail in Chapters 4, 5, and 6.

TABLE 1 Seven Dyes Recognized by Hesse in 1907

Common Name	Later FDA Name[a]
Amaranth	FD&C Red No. 2
Erythrosine	FD&C Red No. 3
Indigo disulfonic acid sodium salt (indigotine)	FD&C Blue No. 2
Light Green SF Yellowish	FD&C Green No. 2[b]
Naphthol Yellow S	FD&C Yellow No. 1[c]
Orange I	FD&C Orange No. 1[d]
Ponceau 3R	FD&C Red No. 1[e]

[a]The FD&C designations were not assigned by the Food and Drug Administration until 1938, when the Food, Drug, and Cosmetic Act was passed (see Chapter 4).
[b]Deleted for food, drug, and cosmetic use in 1966 because it was of insufficient economic importance.
[c]Delisted for food use in 1959, but permitted in externally applied drugs and cosmetics under the name Ext. D&C Yellow No. 7.
[d]Delisted for food use in 1956, but permitted in externally applied drugs and cosmetics under the name Ext. D&C Orange No. 3; delisted completely in 1968.
[e]Delisted for food use in 1961, but permitted in externally applied drugs and cosmetics under the name Ext. D&C Red No. 15; delisted completely in 1966.

TABLE 2 Ten Dyes Added to the Approved List, 1916–1929

Common Name	Later FDA Name	Year Added
Tartrazine	FD&C Yellow No. 5	1916
Sudan I	[a]	1918
Butter Yellow	[a]	1918
Yellow AB	FD&C Yellow No. 3[b]	1918
Yellow OB	FD&C Yellow No. 4[c]	1918
Guinea Green B	FD&C Green No. 1[d]	1922
Fast Green FCF	FD&C Green No. 3	1927
Brilliant Blue FCF	FD&C Blue No. 1	1929
Ponceau SX	FD&C Red No. 4	1929
Sunset Yellow FCF	FD&C Yellow No. 6	1929

[a]Withdrawn after having been on the permitted list for about 6 months because contact dermatitis was observed in up to 90 percent of the factory workers handling these dyes. There were no reports of any harmful effects in those who consumed foods colored with the dyes. Butter Yellow was later found to be carcinogenic in test animals.
[b]Delisted for food use in 1959, but permitted in externally applied drugs and cosmetics under the name Ext. D&C Yellow No. 9; delisted completely in 1960.
[c]Delisted for food use in 1959, but permitted in externally applied drugs and cosmetics under the name Ext. D&C Yellow No. 10; delisted completely in 1960.
[d]Delisted for food, drug, and cosmetic use in 1966 because it was of insufficient economic importance.

3
Functional Use of Color in Food

Impressions of foods are often expressed in terms of characteristic colors: Preferences are for colors associated with maturity in fruits and vegetables, "richness" in gravy or in caramels, certain types and forms of meats, or flavors in certain crop varieties. Food products that appear to lack the expected color value may be thought inferior; thus pink, rather than red, fruit may imply a lack of maturity, very light brown bakery goods may be thought insufficiently baked, or an unnatural color may be taken as evidence of spoilage. Certain colors are well established: Butter, egg noodles, and lemon-flavored cake are expected to be yellow; mint-flavored jelly must be green; and orange drink, imitation or natural, must be orange. Ethnic, geographic, and cultural practices or heritage greatly influence color preferences in certain foods, e.g., egg shells, butter, baked beans, and smoked fish.

Homemakers consider the coordination of colors of food important to successful cookery; in dietetic administration, it is considered important to therapy.

The importance of color to the acceptability of foods by consumers is recognized in many regulatory and market food-grade standards, e.g., color-grade specifications for fruits, vegetables, maple syrup, honey, sugars, oils, beer, and liquors. Factors of importance here include adherence to stipulated color values and uniformity within the product. Consumer preferences for certain color values in foods are sufficiently great that deviation from them markedly affects commercial value.

The preservation, protection, and maintenance of colors in foods are frequently difficult. Natural pigments may well deteriorate because of exposure to light, air, or extremes of temperatures, or because they interact with other constituents of the product. These changes may occur rapidly or over long periods, and may result in fading, darkening, or change in hue, any of which can affect acceptability. Processes or procedures used in preparing foods for preservation or to make them more convenient may alter them chemically and physically and thus affect their color. While many measures are taken to protect and preserve color—such as use of selective packaging materials, precisely controlled processes, and controlled atmospheres—they may be inadequate to prevent color changes from occurring or to produce the colors that are preferred by the consumer.

The addition of coloring materials to foods to achieve preferred color values is a desirable procedure, provided the food is wholesome, safe, and otherwise acceptable. Proper use of color in foods serves several important functions: (a) It helps to correct for natural variations in color or for changes during processing and storage; (b) it makes the food more visually appealing and helps emphasize or identify flavors normally associated with various foods; (c) it assures greater uniformity in appearance, and hence acceptability, by correcting natural variations and irregularities resulting from storage, processing, packaging, and distribution; and (d) it helps preserve the identity or character by which foods are recognized. Misuse of colors to mask inferior quality is wholly unacceptable.

4
Current Usage of Certified Colors in Foods*

The amount of certified color added to any food is self-limiting by virtue of the conditions of good manufacturing practice. Each of the certified colors in use today possesses high tinctorial strength, thus only an extremely small amount is needed to color a product properly. If, for example, too much color is added to gelatin, the product has an "artificial" appearance; if too much is added to an opaque product, the result is a dull shade rather than a desirable bright color. Indeed, a manufacturer is unlikely to use a costly additive in excess when only a little is required to produce the desired effect.

A large part of the food we eat contains no added color. Many foods, including such important staples as bread, fresh meat, potatoes and other vegetables, and most fruits, have an acceptable natural color and are not colored further during processing. Foods that usually require added color include desserts, beverages, candies, and confections. These and other products to which certified colors are added during processing are listed in Table 3, along with the average color concentration and range of concentration for each category. Since only about 10 percent of the food consumed contains added certified color, the color concentration ranges shown in Table 3 must not be taken as representative of the total food intake.

It has been estimated that the concentration of certified colors in the total food supply is about nine parts per million. This value is calculated from the following data: (a) about 1,420 lb of food consumed per capita annually, as reported by the U.S. Department of

*This chapter is based on information contained in the article "Use of Certified FD&C Colors in Food," which appeared in Food Technology 8:946, 1968.

TABLE 3 Major Categories of Processed Food That Use Certified Colors in Their Manufacture and Color Concentration Levels

	Color Concentration	
Category	Range (ppm)	Average (ppm)
Candy and confections	10–400	100
Beverages (liquid and powdered)	5–200	75
Dessert powders	5–600	140
Cereals	200–500	350
Maraschino cherries	100–400	200
Pet foods	100–400	200
Bakery goods	10–500	50
Ice cream and sherbets	10–200	30
Sausage (surface)	40–250	125
Snack foods	25–500	200
Meat stamping inks	—	—
Miscellaneous (salad dressing, nuts, gravy, spices, jams, jellies, food packaging, etc.)	5–400	—

Agriculture; (b) an assumed U.S. population of 200 million; and (c) sales of about 2,540,000 lb of certified food color in the United States in 1967 (estimate based on 1,916,179 lb sold during the first 9 months of 1967, as determined by the Certified Color Industry Committee*).

The amount of each certified color used in the production of each food group for the first 9 months of 1967 is shown in Table 4. From the average concentration of color used for each group (Table 3), and from the poundage of color used in the 9-month period for each group (Table 4), the amount of food produced for each group during the 9-month period can be calculated. The results, presented in Table 5, indicate that a total of 20,821,999,253 lb of foods containing certified colors were produced during the first 9 months of 1967. This represents approximately 10 percent of the total food consumption during this period, or about 104 lb of food per capita.

Values for the daily per capita consumption of food in each category, calculated from the data in Table 5, are used in Table 6 to

*The Certified Color Industry Committee is an informal, unincorporated association of manufacturers of over 90 percent of the certified colors produced in the United States. The Committee was organized to deal with regulatory and legislative problems affecting the entire industry and involving the Food and Drug Administration.

TABLE 4 Food, Drug, and Cosmetic Usage, in lb, of Primary Colors. Figures Represent Sales for the First 9 Months of 1967 and Do Not Include Exports or Sales to Jobbers and Other Manufacturers

Category	FD&C Blue No. 1	FD&C Blue No. 2	FD&C Green No. 3	Orange B	FD&C Red No. 2	FD&C Red No. 3	FD&C Red No. 4	FD&C Violet No. 1	FD&C Yellow No. 5	FD&C Yellow No. 6	Total
Candy, confections	6,632	2,499	124	0	67,637	11,665	0	1,459	59,903	52,770	202,689
Beverages	15,800	2,375	301	0	282,695	1,056	0	985	78,933	181,292	563,437
Dessert powders	3,270	1,659	14	0	62,363	8,616	0	0	59,961	51,622	187,505
Cereals	843	99	0	0	15,558	1,421	0	0	52,496	35,464	105,881
Maraschino cherries	597	0	98	0	8,104	3,469	11,308	0	5,644	4,830	34,050
Pet food	1,473	6,764	0	0	67,058	1,023	0	1,278	101,743	23,226	202,565
Bakery goods	3,680	673	7	0	43,522	9,560	0	369	77,885	42,203	177,899
Ice cream, sherbet, dairy products	2,599	179	7	0	29,697	621	0	45	35,048	23,868	92,064
Sausage	647	0	0	16,890	36,084	4,970	0	0	6,502	99,605	164,698
Snack foods	305	0	0	0	3,623	766	0	2	18,456	11,409	34,561
Meat inks	11	0	0	0	12	10	0	2,223	15	0	2,271
Miscellaneous	5,345	1,990	1,298	0	46,219	18,200	398	1,134	44,841	29,134	148,559
Subtotal (food use)	41,202	16,238	1,849	16,890	662,572	61,377	11,706	7,495	541,427	555,423	1,916,179
Pharmaceutical	3,250	593	220	0	21,179	12,168	1,186	347	17,275	15,938	72,156
Cosmetics	397	30	27	0	3,417	903	630	96	3,125	2,148	10,773
Total	44,849	16,861	2,096	16,890	687,168	74,448	13,522	7,938	561,827	573,509	1,999,108

TABLE 5 Total Production[a] of Processed Foods Using Certified Colors, Calculated and Based on Total Color Sold (Table 4) and the Reported Average Color Concentration (Table 3)

Category	Total Color Sold (lb)	Average Color Concentration (ppm)	Calculated Total Production (lb)
Candy, confections	202,689	100	2,026,890,000
Beverages	563,437	75	7,512,305,521
Dessert powders	187,505	140	1,339,348,215
Cereals	105,881	350	302,502,017
Maraschino cherries	34,050	200	170,250,000
Pet foods	202,565	200	1,012,825,000
Bakery goods	177,899	50	3,557,980,000
Ice cream, sherbet, dairy products	92,064	30	3,038,112,000
Sausage	164,698	125	1,317,584,000
Snack foods	34,561	200	172,805,000
Meat inks	2,271	—	—
Miscellaneous (salad dressing, nuts, gravy, spices, jams, jellies, food packaging, etc.)	148,559	400	371,397,500
Total			20,821,999,253

[a]During first 9 months of 1967.

calculate the extreme upper range of color per category that might be ingested per capita per day, based on the estimated maximum concentration for each food category as shown in Table 3.

Using the production figure for each color in each food category (Table 4), the ratio of individual color to total color is used to estimate the amount of each certified color ingested per capita per day. These data are presented in Table 7.

It is apparent from Table 4 that the largest single user of certified color is the beverage industry. Production figures and estimates of consumption of soft drinks* indicate that the average per capita consumption of beverages for 1965 was about 259 8-oz bottles. Of this amount, approximately 30 percent would contain certified color, since cola beverages, which are colored with caramel, comprise more than 60 percent of total soft drinks consumed and another 10

*Supplied by the American Bottlers of Carbonated Beverages.

TABLE 6 Estimated Total Color That Might Be Ingested per Capita per Day Based on Maximum Color Concentration (Table 3) and Calculated From Total Production[a] (Table 5) for Each Food Category

Category	Total Average Consumption per Capita (lb)	Equivalent (g)	Daily Consumption per Capita (g)	Color Concentration Maximum (ppm)	Maximum Color Ingested per Capita per Day (mg)
Candy, confections	10	4,536	17	400	7
Beverages	38	17,238	64	200	12.8
Dessert powders	6.7	3,039	11	600	6.6
Cereals	1.5	680	3	500	1.5
Maraschino cherries	0.85	386	1	400	0.4
Bakery goods	17.8	8,074	30	500	15
Ice cream, sherbet, dairy products	15	6,806	25	200	5
Sausage	6.6	2,994	11	250	3
Snack foods	0.87	395	2	500	1
Miscellaneous	1.85	839	3	400	1.2
Total					53.5

[a] During first 9 months of 1967.

TABLE 7 Estimated Maximum Amount of Each Color Ingested per Capita, per Day, per Food Category Based on Individual and Total Color Production (Table 4) and Total Color Ingested (Table 6)

Category	Total Color (mg)	FD&C Blue No. 1 (mg)	FD&C Blue No. 2 (mg)	FD&C Green No. 3 (mg)	Orange B (mg)	FD&C Red No. 2 (mg)	FD&C Red No. 3 (mg)	FD&C Red No. 4 (mg)	FD&C Violet No. 1 (mg)	FD&C Yellow No. 5 (mg)	FD&C Yellow No. 6 (mg)
Candy, confections	7	0.23	0.09	0.04	0	2.3	0.40	0	0.05	2.1	1.8
Beverages	12.8	0.36	0.05	0.01	0	6.4	0.02	0	0.02	1.8	4.1
Dessert powders	6.6	0.11	0.06	<0.01	0	2.2	0.30	0	0	2.1	1.8
Cereals	1.5	0.01	<0.01	0	0	0.2	0.02	0	0	0.7	0.5
Maraschino cherries	0.4	0.01	0	0.01	0	0.1	0.04	0.13	0	0.1	0.1
Bakery goods	15	0.31	0.06	0	0	3.7	0.81	0	0.3	6.6	3.6
Ice cream, sherbet, dairy products	5	0.14	0.01	<0.01	0	1.6	0.03	0	<0.01	1.9	1.3
Sausage	3	0.01	0	0	0.31	0.7	0.09	0	0	0.1	1.8
Snack foods	1	0.01	0	0	0	0.1	0.02	0	0	0.5	0.3
Miscellaneous	1.2	0.04	0.02	0.01	0	0.4	0.15	0	0.01	0.4	0.2
Total	53.5	1.23	0.29	0.07	0.31	17.7	1.88	0.13	0.11	16.3	15.5

percent consists of club soda and various lemon and lime drinks that are not colored. Adjusting this figure further to show consumption for 9 months gives an average per capita consumption of 58 8-oz bottles of beverages that are colored with certified colors. The calculated total production figure for beverages (including dry beverage mixes) colored with certified colors shown in Table 5 is 7.51×10^9 lb. On the basis of a 200-million population, the average per capita consumption is 76 8-oz bottles.

The maximum average daily per capita consumption of food colors (53.5 mg, Table 6) represents the amount of color that is equivalent to the largest amount used within the customary range. If the average, rather than the maximum, color concentrations from Table 3 are used in the calculations, the value for the average total color ingested per capita per day (Table 6) is approximately 15 mg rather than 53.5 mg.

5
Food Color Regulations

Regulations promulgated under the Pure Food and Drugs Act of 1906 effected the elimination of undesirable contamination from dyes used in food in the sense that responsible manufacturers voluntarily forwarded each batch of dye to Washington, D.C., for certification that it contained no harmful impurities. In 1938, the Federal Food, Drug, and Cosmetic Act made this certification procedure mandatory, and safety data based on animal tests were supplied to support the continued listing of specified colors under the new Act. Regulations established under this law designated colors permitted for use in food as FD&C colors under Section 406(b), which reads as follows: "The Secretary shall promulgate regulations providing for the listing of coal-tar colors which are harmless and suitable for use in food and for the certification of batches of such colors, with or without harmless diluents."

An application for the admission of a coal-tar color to listing, according to the regulations established under the above section, shall be accompanied by the following requirements:

1. Full reports of investigations that are adequate to show whether such color is harmless and suitable for use in food
2. A full statement of the percentages and compositions of the pure dye and all intermediates and other impurities contained in such color
3. A full statement showing the identity, purity, and quantity or proportion of each intermediate and other article used as a component of such color, and all steps in the process used for the manufacture of such color
4. A full description of practical and accurate methods of analy-

sis for the quantitative determination of the pure dye and of all intermediates and other impurities contained in such color

5. A full description of practical and accurate methods for the identification of such dye in food colored therewith

6. A 5-lb sample of such color (unless the Food and Drug Administration authorizes or requires submission of another quantity suitable to the need for investigation) taken from a batch produced under practical manufacturing conditions, and accurately representative of such batch

7. The advance monetary deposit prescribed.

The above conditions for listing a new color are similar to those required for approval of a food additive under the Food Additives Amendment of 1958. (See Tables 8 and 9 for the food colors recognized under the 1938 act).

Lakes of the water-soluble FD&C colors, which are prepared by extending the aluminum or calcium salt of the color on a substratum of alumina, have been permitted for use in coloring nonaqueous food products since 1959. Each lake is considered to be a straight color and is listed under the name of the color from which it is formed.

It should be noted that two expressions in Section 406(b) of the law, "coal-tar colors" and "harmless," led to certain problems in the certified color industry. Originally, the term "coal-tar colors" was applied to this group of synthetic food colors because coal-tar was the source of chemicals from which they were derived. Con-

TABLE 8 Fifteen Coal-Tar Colors Recognized for Food Use under the 1938 Act

FD&C Blue No. 1 (Brilliant Blue FCF)
FD&C Blue No. 2 (sodium salt of indigo disulfonic acid)
FD&C Green No. 1 (Guinea Green B)
FD&C Green No. 2 (Light Green Yellowish)
FD&C Green No. 3 (Fast Green FCF)
FD&C Orange No. 1 (Orange I)
FD&C Red No. 1 (Ponceau 3R)
FD&C Red No. 2 (Amaranth)
FD&C Red No. 3 (Erythrosine)
FD&C Red No. 4 (Ponceau SX)
FD&C Yellow No. 1 (Naphthol Yellow S)
FD&C Yellow No. 3 (Yellow AB)
FD&C Yellow No. 4 (Yellow OB)
FD&C Yellow No. 5 (Tartrazine)
FD&C Yellow No. 6 (Sunset Yellow FCF)

TABLE 9 Four Colors Added to the List of Permitted Colors between 1939 and 1950

Common Name	FDA Name	Year Added
Naphthol Yellow S, potassium salt	FD&C Yellow No. 2[a]	1939
Oil Red XO	FD&C Red No. 32[b]	1939
Orange SS	FD&C Orange No. 2[c]	1939
Benzyl Violet 4B	FD&C Violet No. 1	1950

[a]Delisted for food use in 1959, but permitted in externally applied drugs and cosmetics under the name Ext. D&C Yellow No. 8; delisted completely in 1962.
[b]Delisted for food use in 1956, but permitted in externally applied drugs and cosmetics under the name Ext. D&C Red No. 14; delisted completely in 1963.
[c]Delisted for food use in 1956, but permitted in externally applied drugs and cosmetics under the name Ext. D&C Orange No. 4; delisted completely in 1963.

siderable unfavorable publicity has stemmed from the association of food colors with coal-tar, the popular conception of which is of a thick, black, sticky substance. In fact, of course, the derivatives from coal-tar are isolated and highly purified prior to use in making food colors. The food colors themselves are specific chemical entities, distinct from the products from which they are derived. Later on it was found that identical starting materials could be obtained from other sources, for example, petroleum.

The term "harmless" was brought into focus as a result of publicity associated with a special committee of the House of Representatives formed in 1950 to investigate the use of chemicals in and on food products. The FD&C colors were the only group of food additives certified by FDA as harmless. Accordingly, FDA initiated an animal testing program that required more animals and higher feeding levels than did the earlier pharmacological studies (1938–1940).

Furthermore, FDA redefined "harmless" to mean a substance incapable of producing harm in test animals in any quantity or under any conditions. Under this interpretation, FDA delisted eight FD&C colors between 1956 and 1960 (see Table 12, Chapter 6). Realizing that full implementation of this interpretation would lead to delisting of practically all colors, the Department of Health, Education, and Welfare proposed remedial legislation in 1960,* intended to give

*In March 1959, Congress amended the Food, Drug, and Cosmetic Act of 1938 to permit temporary listing and certification of Citrus Red No. 2 (to replace the delisted FD&C Red No. 32) for coloring the skins of oranges not intended for processing.

FDA the authority to establish specifically quantitative limits on the use of certified colors, thus defining their safe application and assuring their continued use on a stable and realistic basis. The Certified Color Industry Committee worked closely with the Department and FDA in seeking the legislation, with the result that on July 12, 1960, the Color Additive Amendments of 1960 became Public Law 86-618.

The Color Additive Amendments of 1960 made the following major changes in the Federal Food, Drug, and Cosmetic Act: (a) FDA was explicitly authorized to set safe limits, or tolerances, on the amount of color permitted in foods, drugs, and cosmetics. Previously, a color that could be shown to produce injury to test animals, in any amount, no matter how large, could be ruled out entirely, even though its use in small amounts under specified conditions was entirely safe; (b) all colors for foods, drugs, and cosmetics were brought under the premarketing safety clearance provisions. Previously, these provisions had applied only to the so-called "coal-tar" colors, not to the "natural" colors; (c) FDA was authorized to require that previously authorized colors be retested by the manufacturer for safety, using modern techniques and procedures, where any question of safety may have arisen since the original listing of the color as safe.

On June 22, 1963, regulations under the new law were published in the Federal Register. These regulations outline for manufacturers the type of experimental data and other information required and how the information should be submitted, to obtain safety clearance for permanent listing and for setting safe tolerances for color additives. The regulations cover such matters as definition of terms, fees to be charged for listing and certification of batches of colors, labeling requirements for colors, time schedules for acting upon petitions, protection of trade secrets, procedures for obtaining certification or exemption from certification of batches of both synthetic and natural colors, and procedures for filing objections and requesting public hearings on regulations.

Safety data that may be required under the regulations include detailed data from appropriate animal and other biological experiments; information on chemical identity and composition and physical, chemical, and biological properties; a description of tests, facilities, and controls used in manufacture; data on stability, including a proposed expiration date where appropriate; and, when needed, satisfactory methods for detecting and measuring the color in the products in which it would be used.

The regulations provide that a safety factor of 100 to 1 will ordinarily be used in extrapolating "no-adverse-effect levels" in animals to safe levels for man unless use of a different factor is supported by the data submitted; also they take into account any probably additive effect of the toxicity of the color in question with that of other related colors or with food additives or pesticides that may also be present in foods.

Where the data submitted do not establish safety for all uses of the color proposed, the new color law allows FDA to make allocations among competing needs. The regulations require the submission of data by all interested parties before allocations are made.

Provision is made for referral to an Advisory Committee upon request of the sponsor, whenever the Commissioner of Food and Drugs believes the data establish that the color additive is carcinogenic and thus could not be permitted in any amount under the anticancer clause of the law. The commissioner may upon his own initiative refer the matter to an adivsory committee under similar circumstances. The advisory committee, once formed, would be requested to study the data and report its conclusions and recommendations to the commissioner within 60 days, unless an extension of time is authorized upon request of the committee.

Public Law 86-618 also made possible, through provisional listing, the continued use of commercially established color additives to an extent consistent with public health, pending completion of scientific tests required for definitive listing. On January 11, 1963, the FDA commissioner published a revision to the Color Additive Regulations, extending the provisional listing of the FD&C colors until June 1, 1964, for some, and until August and October 1964 for others. Subsequent revisions extended the closing date to December 31, 1971. During this period, the animal testing program was to be completed by FDA's Division of Pharmacology, and methods of analysis were to be collaboratively studied by manufacturers.

In mid-1969, the Food and Drug Administration issued regulations permitting the permanent listing of FD&C Blue No. 1, FD&C Red No. 3, and FD&C Yellow No. 5 for use in coloring foods generally in amounts consistent with good manufacturing practice; FD&C Red No. 40 was added to the list in April 1971. Citrus Red No. 2 had been earlier listed under the restriction that it be used only for coloring the skins of oranges that are not intended or used for processing, at a level of not more than 2 ppm; Orange B was listed in 1966 under the restriction that it be used only for coloring the casings or surfaces of frankfurters and sausages, at a level of not more than 150 ppm.

6 Studies by International Agencies

Two meetings that dealt primarily with food colors have been held by the Joint FAO/WHO Expert Committee on Food Additives (1963, 1965): one in Rome, Italy, in December 1969, and one in Geneva, Switzerland, in December 1964. In both instances, the purpose was to evaluate the toxicological hazards attending the use of food colors and to establish specifications of identity and purity for those colors.

Food colors were also considered, but to a lesser extent, by the Expert Committee (1966; 1967; 1970a, b, c) at meetings in 1966 and 1969. In all, over 160 food colors that had been used in various countries were studied and placed by the Expert Committee, in accordance with its toxicological evaluation, in the following categories:

Category A—Colors found acceptable for use in food. For each of these colors, a maximum acceptable daily intake value was established. The Expert Committee emphasized that the assignment of a color to this category should not be interpreted as indicating that further research was unnecessary. Indeed, the Committee recognized that more work was called for, especially in view of the possibility that research might well provide more precise means of toxicological assessment; additional investigations of the effects on reproduction and the fetus are particularly needed.

Category B—Colors for which the data available to the Expert Committee were not wholly sufficient to meet the requirements of Category A.

Category C I—Colors for which the available data were inadequate for evaluation, but for which a substantial amount of detailed information from results of long-term tests was available.

Category C II—Colors for which the available data were inadequate for evaluation and for which virtually no information on long-term toxicity was available. Colors on which there were data from long-term tests for tumor formation unaccompanied by information from other long-term studies were considered as falling within this category.

Category C III—Colors for which the available data were inadequate for evaluation, but which indicated the possibility of harmful effects.

Category D—Colors for which virtually no toxicological data were available.

Category E—Colors found to be harmful and that should not be used in food.

On the basis of available analytical information, the Committee also classified colors from the standpoint of chemical specifications as follows:

I. Food colors for which the Committee was able to prepare satisfactory specifications.

II. Food colors for which the Committee chose not to prepare specifications, since toxicological studies on substances of defined composition were in progress.

III. Food colors for which the chemical information available to the Committee was inadequate to permit the preparation of fully satisfactory specifications.

IV. Food colors for which the Committee did not attempt to prepare specifications either because toxicological data were totally lacking or because the colors were demonstrably toxic at levels that indicated that their use in foods is undesirable.

Some food colors now used in the United States have been classified by the Expert Committee as shown in Table 10. Acceptable daily intakes established for colors classified in toxicological Category A are given in Table 11.

TABLE 10 Classification of Some U.S. Food Colors by the Joint FAO/WHO Expert Committee on Food Additives

Common Name	FD&C No.	Color Index No.[a]	Classification	
			Specification	Toxicological
Amaranth	Red No. 2	16185	I	A
Annatto, bixin, and norbixin		75120	I	A
β-Apo-8′-carotenal			I	A
β-Apo-8′-carotenoic acid, methyl or ethyl ester			I	A
Beet red and betanin			III	b
Benzyl Violet 4B	Violet No. 1[c]	42640	I	C III
Brilliant Blue FCF	Blue No. 1	42090	I	A
Canthaxanthine			I	A
Caramel		Included but not numbered	IV	D
Carbon black		77266	II	b
Carotene (natural)		75130	III	b
β-Carotene (synthetic)			I	A
Citrus Red No. 2	Citrus Red No. 2[d]	12156	I	E
Cochineal and carminic acid		75470	III	b
Erythrosine	Red No. 3	45430	I	A
Fast Green FCF	Green No. 3	42053	I	A
Indigotine	Blue No. 2	73015	I	A
Iron oxides		77489 77491 77492 77499	III	b

TABLE 10
Continued

Common Name	FD&C No.	Color Index No.[a]	Classification	
			Specification	Toxicological
Ponceau SX	Red No. 4[e]	14700	IV	E
Riboflavin			I	A
Saffron, crocin, and crocetin		75100	III	[b]
Sunset Yellow FCF	Yellow No. 6	15985	I	A
Tartrazine	Yellow No. 5	19140	I	A
Titanium dioxide		77891	I	A
Turmeric (curcumin)		75300	I	A
Ultramarine		77007 77013	III	[b]
Xanthophylls			III	[b]

[a]The color index number is that number assigned to a specific color listed in the _Color Index_, 2nd Edition, 1956, compiled by the Society of Dyers and Colorists and by the American Association of Textile Chemists and Colorists.
[b]No attempt was made at a toxicological evaluation because of the lack of knowledge of composition and the paucity of biological studies.
[c]A petition for the permanent listing in the United States is under study by the FDA.
[d]Use in the United States is limited to the coloring of skins of oranges not intended or used for further processing, at a level not exceeding 2 ppm by weight.
[e]Use in the United States is limited to the coloring of maraschino cherries, at a level not exceeding 150 ppm by weight.

TABLE 11 Acceptable Daily Intakes (ADI) for Some U.S. Food Colors as Established by the FAO/WHO Expert Committee on Food Additives

Common Name	FD&C No.	Color Index No.	Acceptable Daily Intake for Man (mg per kg body weight)		
			Unconditional[a]	Conditional[b]	Temporary[c]
Amaranth	Red No. 2	16185	0– 1.5		
Annatto, bixin, and norbixin		75120			0–1.25[d]
β-Apo-8′-carotenal			0– 2.5[e]	2.5– 5.0[e]	
β-Apo-8′-carotenoic acid, methyl or ethyl ester			0– 2.5[e]	2.5– 5.0[e]	
Brilliant Blue FCF	Blue No. 1	42090	0–12.5		
Canthaxanthin			0–12.5	12.5–25.0	
β-Carotene (synthetic)			0– 2.5[e]	2.5– 5.0[e]	
Erythrosine	Red No. 3	45430			0–1.25[f]
Fast Green FCF	Green No. 3	42053	0–12.5		
Indigotine	Blue No. 2	73015			0–2.5[g]
Riboflavin			0– 0.5		
Sunset Yellow FCF	Yellow No. 6	15985	0– 5.0		

Tartrazine	Yellow No. 5	19140	0– 7.5		
Titanium dioxide		77891	[h]		
Turmeric (curcumin)		75300			0–0.5[i]

[a]An unconditional ADI was allocated only to those substances for which the biological data available included either the results of adequate short-term and long-term toxicological investigations or information on the biochemistry and metabolic fate of the compound or both.

[b]A conditional ADI was allocated for specific purposes arising from special dietary requirements.

[c]A temporary ADI was allocated when the available data were not fully adequate to establish the safety of the substance, and it was considered necessary that additional evidence be provided within a stated period of time. If the further data requested do not become available within the stated period (see footnotes d, f, g, and i), it is possible that the temporary ADI will be withdrawn by a future Committee.

[d]Further work required by June 1972: metabolic studies on the major carotenoids of annatto.

[e]Expressed as total carotenoids by weight.

[f]Further work required by June 1972: studies on the metabolism in several species and preferably in man and elucidation of the mechanism underlying the effect of this color on plasma-bound iodine levels.

[g]Further work required by June 1974: 2-year study in a nonrodent mammalian species.

[h]Not limited except by good manufacturing practice.

[i]Further work required by June 1974: studies on the metabolism of curcumin and a 2-year study in a nonrodent mammalian species.

7
Safety of Food Colors

In 1938, with the enactment of the Federal Food, Drug, and Cosmetic Act, provision was made for the continued use of "harmless" synthetic colors. In 1939, after public hearings, 18 food colors were so designated. From 1938 to 1940, limited pharmacological testing of certain food colors was conducted by the Food and Drug Administration, but the need for further studies was not apparent until an incident occurred in which some children were made ill by candy colored with a permitted orange color. As it happened, the substance in question had been added to the candy batch without proper dilution, resulting in a very high concentration of color. Because of this and similar occurrences, FDA began a new testing program. On the basis of their findings, seven food colors were delisted and action was started to delist a number of others. This action initiated a long chain of events leading to the passage of the Color Additives Amendment in 1960 and a requirement that long-term dietary studies be made in two or more species on all colors to be certified for use in food.

The new color additives law went into effect on July 12, 1960. It made provision for the continued use of commercially established color additives through provisional listings. Since then, the closing dates for provisional listing of food colors have been extended a number of times to permit completion of the required animal studies, adequate evaluation of the resulting data, and preparation of petitions requesting the permanent listing as called for in Section 706(b)(1) of the Act. The long-term animal feeding studies, most of which were performed by the Toxicology Branch of FDA and paid for from funds derived from certification fees paid by color manufacturers, have been completed, and the results have been published or reported informally. Data on color additives formerly certified for food use, but delisted on the basis of toxicological evaluation, are summarized in Table 12.

Petitions requesting the permanent listing of all certified food colors for which "no-adverse-effect" levels in animal feeding studies were established have been submitted to the Food and Drug Administration by the Certified Color Industry Committee.

The available toxicological data for provisionally or permanently listed synthetic food colors subject to certification are shown in

TABLE 12 Colors Formerly Certified for Use in Food in the United States That Have Been Delisted on the Basis of Toxicological Evaluation[a]

Color	Species	Dose Level Percent in Diet	No-Adverse-Effect Level (%)
FD&C Orange No. 1	Rat	0.1 –2	0.1
	Dog	0.02–1	0.02
FD&C Orange No. 2	Rat	0.01–0.25	Questionable at 0.01
	Dog	0.02–0.2	0.02
FD&C Red No. 1	Rat	0.5 –5	Effect at all levels in all species studied
	Rat	2	
	Dog	0.25–2	
	Mouse	1 and 2	
	Mouse	0.1 and 0.5	
FD&C Red No. 32	Rat	0.1 and 0.25	Not established
	Rat	0.001 and 0.01	0.01
	Dog	0.01–0.2	Not established
FD&C Yellow No. 1[b]	Rat	0.1 –2	Not established
	Dog	0.02–0.4	0.4
FD&C Yellow No. 2 (potassium salt of FD&C Yellow No. 1)	—	—	Same biological conclusions apply as for FD&C Yellow No. 1
FD&C Yellow No. 3	Rat	0.05–0.25	0.05
	Dog	0.05–0.5	0.05
FD&C Yellow No. 4	Rat	0.05–0.25	0.05
	Dog	0.05–0.5	Not established

[a]With the exception of FD&C Red No. 1, all colors in this table were delisted under the "harmless per se" principle prior to passage of the Color Additives Amendment in 1960. Feeding tests with FD&C Red No. 1 demonstrated it to be toxic upon ingestion, and provisional listing of the color was terminated in November 1960.

[b]Although this color has been delisted for use in food, it is permitted for use in externally applied drugs and cosmetics under the name Ext. D&C Yellow No. 7.

Table 13, which also gives the acceptable daily intakes for certain of the colors that have been established by the Joint FAO/WHO Expert Committee on Food Additives. It should be noted, however, that the Expert Committee, at its 1969 meeting in Rome, Italy, reconsidered the significance of subcutaneous sarcoma as related to safety of ingested colors and recognized that an "advance has taken place in our understanding of the mechanism of a pathological process, namely the development of subcutaneous sarcoma in rats and mice, at the site of repeated injection of food colourings and other additives. This development has made possible a reevaluation of colourings such as Brilliant Blue FCF (FD&C Blue No. 1) and Fast Green FCF (FD&C Green No. 3), previously placed in Category B." Category B, as explained in Chapter 5, includes those colors for which the toxicological data available to the Expert Committee were not sufficient to justify their being judged entirely acceptable for food use. The two colors mentioned were originally placed in Category B because they were reported to produce a significant incidence of sarcoma at the site of repeated subcutaneous injections. When the colors were re-evaluated in 1969, they were placed in toxicological Category A (acceptable for food use), and acceptable daily intakes were established for them.

The requirements for safety for food colors are the same as for the safety of other food additives; they include scientific judgment and evaluation. Some of the newer techniques being suggested for detecting carcinogenicity, mutagenicity, or teratogenicity are not included in all cases, since some of the studies were initiated several years ago (see Chapter 4).

The procedures used by FDA to evaluate the safety of food colors were outlined by W. H. Hansen (1962). Subacute or short-term feeding studies on rats and dogs were conducted for a period of 3 months to determine the maximum tolerable dose and also the minimum grossly toxic dose. These studies were, in turn, used to determine dosage levels for the examination of chronic effects. These latter studies, performed with two or three species, were designed with two questions in mind: What is the maximum no-adverse-effect level for the most susceptible experimental animal tested? What dosage level produces an effect either on the whole animal or on one or more vital organs? All animals in the chronic study were maintained on experimental or control diets for a minimum of 2 years; if the test diet indicated a carcinogenic potential, trials on dogs were continued for a total of 7 years. During the chronic studies, records were kept of mortalities, gross abnormalities, weekly body weights, and food consumption. Complete blood counts were performed at four intervals. At termination of tests, survivors were killed and necropsied: Organ weights and gross pathology were recorded; and tissues taken

TABLE 13 Data on Certified Food Colors That Have Been Provisionally or Permanently Listed in the United States

Color	No-Adverse-Effect Dietary Levels Animal Studies[a]	Safe Level for Man (mg/day)[b]	Estimated Maximum Ingestion (mg/day/capita) (1)	Acceptable Daily Intake (mg/kg) FAO/WHO[a]
FD&C Blue No. 1	5.0% Rats 2.0% Dogs (2,3)	363	1.23	0–12.5 (4)
FD&C Blue No. 2	1.0% Rats — Dogs[d] (c,3)	181	0.29	0– 2.5 (4,e)
FD&C Green No. 1	0.5% Rats — Dogs[g] 2.0% Mice (f)	91	Not applicable[h]	Not established
FD&C Green No. 2	5.0% Rats 1.0% Dogs (f)	181	Not applicable[h]	Not established
FD&C Green No. 3	5.0% Rats 1.0% Dogs 2.0% Mice (c)	181	0.07	0–12.5 (4)
Orange B	5.0% Rats 1.0% Dogs 5.0% Mice (c)	181	0.31	Not evaluated
Citrus Red No. 2	0.1% Rats (5)	18	Not applicable[i]	Not applicable
FD&C Red No. 2	2.0% Rats 2.0% Dogs 2.0% Mice (g)	363[m]	17.7	0– 1.5 (6)
FD&C Red No. 3	0.5% Rats 2.0% Dogs 2.0% Mice (c)	91	1.88	0– 1.25 (4,e)
FD&C Red No. 4	5.0% Rats 1.0% Dogs (7,k)	181	0.13 (j)	Not established
FD&C Violet No. 1	0.5% Rats — Dogs[l] (c)	91	0.11	Not established
FD&C Yellow No. 5	2.0% Rats 2.0% Dogs (c,2)	363	16.3	0– 7.5 (6)

TABLE 13
Continued

Color	No-Adverse-Effect Dietary Levels Animal Studies[a]	Safe Level for Man (mg/day)[b]	Estimated Maximum Ingestion (mg/day/capita) (1)	Acceptable Daily Intake (mg/kg) FAO/WHO[a]
FD&C Yellow No. 6	2.0% Rats 2.0% Dogs 2.0% Mice ([c])	363	15.5	0– 5.0 (6)

[a]Numbers in parentheses refer to references below:
(1) Certified Color Industry Committee, 1968. (See also Table 7, Chapter 3.)
(2) Davis et al., 1964.
(3) Hansen et al., 1966.
(4) Joint FAO/WHO Expert Committee on Food Additives, 1970.
(5) Paynter, O. E., and R. A. Scala, unpublished report.
(6) Joint FAO/WHO Expert Committee on Food Additives, 1967.
(7) Davis et al., 1966.

[b]Safe level for human ingestion is calculated to be 1/100 of the maximum no-adverse-effect level established by long-term animal feeding studies (for the most sensitive species), and assuming a daily dietary intake of 1814 g (U.S. Dept. Agr.) for humans.

[c]Information contained in Color Additive Petition submitted to the Food and Drug Administration by the Certified Color Industry Committee, 1965–1968.

[d]Feeding study at levels of 1 and 2 percent failed to establish a no-adverse-effect level because of deaths from intercurrent virus infections; however, no clinical, gross, or microscopic pathology was attributed to ingestion of FD&C Blue No. 2.

[e]Temporary acceptable daily intake.

[f]Information from an unpublished tabulation (prepared by Hazleton Laboratories, Inc., August 2, 1963) of Food and Drug Administration reports, meetings, and publications regarding safety evaluation studies of FD&C colors performed in the laboratories of the Food and Drug Administration.

[g]Effect of 1 and 2 percent uncertain because of limited number of animals.

[h]FD&C Green No. 1 and No. 2 were deleted and certificates canceled as of October 4, 1966 (30 FR 83333, April 7, 1966), because they were of insufficient economic importance.

[i]Citrus Red No. 2 is used only for coloring the skins of oranges not intended for processing, and its level of use is restricted to 2.0 ppm calculated on the weight of the whole fruit.

[j]Provisional listing of FD&C Red No. 4 for use in food limits the use to coloring maraschino cherries only at a level not to exceed 150 ppm by weight of the cherries. (Drug use is permitted provided that no more than 5.0 mg of the color is ingested per day in drugs prescribed for not longer than 6 weeks' administration.)

[k]Details of dietary feeding to dogs (0.5–2.0 percent) are not yet available.

[l]Information contained in Color Additive Petition submitted to the Food and Drug Administration by the Pharmaceutical Manufacturers Association, March 27, 1968.

[m]The status of FD&C Red No. 2 is under review.

for microscopic examination. Statistical evaluations were made of growth, hematologic, and organ weight data.

The Color Additives Amendment of 1960 brought all food colors (and those for drugs and cosmetics) under premarketing safety clearance provisions. Previously only the synthetic colors were so regulated. A number of previously noncertified natural colorants in commercial use prior to enactment of the new law were provisionally listed and subjected to the same requirements for permanent listing as were the certified synthetic organic colors. Virtually all of these colors have been permanently listed as a result of petitions submitted to FDA or, alternatively, when the commissioner of FDA on his own initiative proposed the listings. Noncertified color additives for which regulations have been issued are listed in Table 14. Carbon black remained on the provisional list for food use as of December 31, 1971.

At its earlier meetings, the Joint FAO/WHO Expert Committee on Food Additives considered the available information on natural colors, except for canthaxanthin and the synthetic carotenes, to be inadequate for evaluation. However, at the 1969 meeting, several natural and noncertified colors were re-evaluated on the basis of the additional information that had become available, and acceptable dietary intakes were established for these colors. Table 15 gives the FAO/WHO acceptable dietary intakes for the colors, together with the Food and Drug Administration's no-adverse-effect levels obtained from animal studies and the calculated safe level for man.

TABLE 14 Color Additives Permanently Listed for Food Use, Exempt from Certification

Color	Use Limitation[a]
Algae meal, dried	For use in chicken feed to enhance the yellow color of chicken skin and eggs
Annatto extract	
β-Apo-8′-carotenal	Not to exceed 15 mg/lb, or pint, of food
Beets, dehydrated (beet powder)	
Canthaxanthin	Not to exceed 30 mg/lb, or pint, of food
Caramel	
β-Carotene	
Carrot oil	
Cochineal extract; carmine	
Corn endosperm oil	For use in chicken feed to enhance the yellow color of chicken skin and eggs
Cottonseed flour, partially defatted, cooked, toasted	
Ferrous gluconate	For coloring ripe olives
Fruit juice	
Grape skin extract	For coloring beverages
Iron oxide (synthetic)	For coloring pet food, not to exceed 0.25 percent by weight of the food
Paprika	
Paprika oleoresin	
Riboflavin	
Saffron	
Tagetes meal and extract (aztec marigold)	For use in chicken feed to enhance the yellow color of chicken skin and eggs
Titanium dioxide	Not to exceed 1 percent by weight of the food
Turmeric	
Turmeric oleoresin	
Ultramarine blue	For coloring salt intended for animal feed, not to exceed 0.5 percent by weight of the salt
Vegetable juice	

[a]Unless otherwise indicated, the color may be used for the coloring of food generally in amounts consistent with good manufacturing practice.

TABLE 15 No-Adverse-Effect Levels and Acceptable Daily Intakes for Some Noncertified Colors

Color	No-Adverse-Effect Dietary Level (%)[a]	FAO/WHO Acceptable Daily Intake for Man (mg per kg body weight)	
		Unconditional[b]	Conditional[b]
Annatto, bixin, and norbixin	0.5	0– 1.25[c]	
β-Apo-8′-carotenal	0.5	0– 2.5[d]	2.5– 5.0[d]
β-Apo-carotenoic acid, methyl or ethyl ester[e]	1.0	0– 2.5[d]	2.5– 5.0[d]
Canthaxanthin	5.0	0–12.5	12.5–25.0
β-Carotene (synthetic)	0.1	0– 2.5[d]	2.5– 5.0[d]
Riboflavin	—	0– 0.5	
Titanium dioxide	—	[f]	
Turmeric (curcumin)	—	0– 0.5[c]	

[a]Established by long-term (2 years) dietary feeding and three- to four-generation reproduction studies in rats.
[b]For explanation of "unconditional," "conditional," and "temporary" ADI's, see footnotes a, b, and c to Table 11, Chapter 5.
[c]Temporary ADI established at 1969 meeting of Expert Committee.
[d]Expressed as total carotenoids by weight.
[e]This color is not on FDA's proposed or permanent list of noncertified colors.
[f]Not limited except as by good manufacturing practice.

8
Physical and Chemical Properties Desired in Food Colors

The chemical and physical properties desired in a food coloring material are few and simple. They are usually not, however, easy to achieve.

The following criteria are listed in a rough order of importance. The first two, concerning safety and flavor, must be fully satisfied in all cases; the remaining are seldom completely met. While the listed order of importance is generally valid, a color that is seriously deficient in any respect may be ruled out for particular applications.

1. Food colors must be safe for human beings at levels that can reasonably be expected to be consumed when they are properly used for the purpose intended.
2. At the level used, the color must either be tasteless and odorless (as with the certified synthetics), or its organoleptic properties must be inoffensive and must blend well with those of the food it colors (as with paprika, saffron, and turmeric).
3. A color should be stable under the influence of: (a) light, (b) oxidation and reduction, (c) pH change, (d) microbiological attack.
4. A color should be compatible with other food components.
5. A color should have high tinctorial strength, usually revealed as an inverse function of the range of concentrations required to color food.
6. A color should possess a desirable hue range. This implies (a) available colors sufficient to cover all useful hues without excessive need for blending, and (b) colors whose absorption characteristics give them high _chroma_ (saturation) and _value_ (brightness),

i.e., strong colors that contain little black (minimum general absorption.

7. A color should be highly soluble in water and in other inexpensive acceptable polar food grade solvents (e.g., alcohol, propylene glycol).

8. Solubility in edible fats and oils is a desirable property, particularly for certain hues.

9. A color should be easily dispersible, if it is not soluble.

10. A color should be inexpensive in terms of the cost to achieve the desired color level.

Tables 16 and 17 describe the performance of all certified colors and most colors exempt from certification currently listed or proposed for use in the United States. The principal conclusions that may be drawn follow:

- There is a need for a stable, intense, true red for general use. Only FD&C Blue No. 1 is a satisfactory blue. There is only one green. Although FD&C Yellow No. 6 (Sunset Yellow) provides a reddish yellow, there are no oranges for general use. It is possible to meet the need, in part, by blending available colors, but this is often a less than satisfactory solution.
- When two colors are blended, the result is duller than either of the originals. Since both components retain their original light absorption, this more general absorption reduces value (i.e., introduces more black). It also produces a color that has less saturation or chroma (i.e., is weaker) than either of its components. The farther apart in hue the components are that are used for blending, the less satisfactory the results. More "primary" hues that would fill some of these needs would permit substantially more effective blending.
- Colors with high saturation and value are highly desirable since the food to which they will be added usually has some color or grayness of its own, the effect of which is to lower saturation and value. Finally, blended colors frequently introduce other disadvantages; e.g., the components may distribute unequally between different food phases or ingredients, and their responses to change in pH and their chemical reactivity may differ.
- No certified oil-soluble colors are available for general use. There are only a few oil-soluble colors, and their hue range and stability are limited. Lakes (insoluble salts) of the certified colors are available and for most purposes are a fairly satisfactory substitute. However, they tend to lack intensity and to disperse poorly.

TABLE 16 Physical and Chemical Properties of Certified Food Colors

FD&C Name (Common Name)	Chemical Class	Stability to			Compatability w/Food Components	Tinctorial Strength	Hue	Solubility (g/100 ml)					Overall Performance
		Light	Oxidation	pH Change				Water	25% EtOH	Glycerin	Prop Gly[a]	Veg Oil	
Red No. 2 (amaranth)	Monoazo	mod	fair	good	good	good	bluish red	20	7	18	1	insol	mod
Red No. 3 (erythrosine)	Xanthine	fair	fair	poor	poor	v good	bluish pink	9	8	20	20	insol	poor–good
Red No. 4[b] (Ponceau SX)	Monoazo	v good	fair	good	v good	v good	yellowish red	11	1.4	5.8	2	insol	good
Red No. 40	Monoazo	v good	fair	good	v good	v good	yellowish red	25	9.5	3	1.5	insol	good
Citrus Red No. 2[c]	Monoazo					good	orangish red	insol	v sl sol	v sl sol	v sl sol		
Orange B[d]	Monoazo	mod	fair			good	orangish yellow						
Yellow No. 6 (Sunset Yellow FCF)	Monoazo	mod	fair	good	mod	good	reddish	19	10	20	2.2	insol	mod
Yellow No. 5 (tartrazine)	Pyrazolone	good	fair	good	mod	good	lemon yellow	20	12	18	7	insol	good
Green No. 3 (Fast Green FCF)	TPM[e]	fair	poor	good	good	exc	bluish green	20	20	20	20	insol	good
Blue No. 1 (Brilliant Blue FCF)	TPM[e]	fair	poor	good (unstable in alkali)	good	exc	greenish blue	20	20	20	20	insol	good
Blue No. 2 (indigotine)	Indigoid	v poor	poor	poor	v poor	poor	deep blue	1.6	0.5	1	0.1	insol	v poor
Violet No. 1 (Benzyl Violet 4B)	TPM[e]	fair	poor	good	fair	exc	bright violet	20	20	20	20	insol	mod

[a]Prop Gly = propylene glycol.
[b]Food use restricted to coloring maraschino cherries at a level not to exceed 150 ppm by weight.
[c]Food use restricted to coloring skins of oranges not intended or used for processing at a level not to exceed 2 ppm by weight.
[d]Food use restricted to coloring the casings or surfaces of frankfurters and sausages at a level not to exceed 150 ppm by weight.
[e]TPM = triphenylmethane.

TABLE 17 Physical and Chemical Properties of Some Noncertified Food Colors

FDA Name (Chemical Class)	Stability to Light	Oxidation	pH Change	Microb Attack	Compatability w/Food Components	Tinctorial Strength (Effective Conc. in Food)	Hue Range	Solubility (g/100 ml) Water	25% EtOH	Glycerin	Prop Gly[a]	Veg Oil	Overall Performance
Annatto extract (carotenoid)	mod	v good	v good		v good	good (0.5–10 ppm)	yellow-peach					0.1	good
β-Apo-8′-carotenal[b] (carotenoid)	fair	poor	good	poor	good	good (1–20 ppm)	light to dark orange	insol[c]	insol	insol	insol		mod
Beets, dehydrated [beet powder] (anthocyanin)	good	exc	good	good	exc	good (1,000–5,000 ppm)	bluish red <pH 6; brown > pH 6						good
Canthaxanthin[d] (xanthine)	mod	fair	good		good	good (5–60 ppm)	pink to red	insol	insol		trace	0.02	mod
Caramel	good	good	good	fair	good	fair (1,000–5,000 ppm)	yellowish tan to red-brown	∞	∞			insol	good
β-Carotene (carotenoid)	fair	poor	good	poor	good	good (2.5–50 ppm)	yellow in oil; orange in water	insol[c]	insol	insol	insol	0.1	mod
Carrot oil (Carotenoid)													
Cochineal extract; carmine (anthroquinone)	good	good	poor	poor	good	mod (25–1,000 ppm)	orange-red → wine-red acid → base	insol[c]	insol	insol	insol	insol	good
Cottonseed flour, partially defatted, cooked, toasted	good	good	good	good		poor (1,000–20,000 ppm)	light to dark brown	insol[c]	insol	insol	insol	insol	
Ferrous gluconate[b]								sol	insol				
Fruit and vegetable juice	poor	poor	v poor–good	poor	poor–good	poor (0.5–5%)		∞	∞	∞	∞	insol	fair
Grape skin extract[e]	poor	poor	poor	mod	good	poor (0.5–5%) (15% color)	red → blue → green > pH →	∞	∞			insol	

Continued overleaf

TABLE 17
Continued

FDA Name (Chemical Class)	Stability to: Light	Oxidation	pH Change	Microb Attack	Compatability w/Food Components	Tinctorial Strength (Effective Conc. in Food)	Hue Range	Solubility (g/100 ml): Water	25% EtOH	Glycerin	Prop Gly[a]	Veg Oil	Overall Performance
Paprika extract and oleoresin (carotenoid)	poor	poor	good	fair	good	good (0.2–100 ppm)	orange to bright red	insol[c]	sol	insol	v sl sol	∞	fair
Riboflavin													
Saffron (isoprenoid)	exc	good	good	v good	v good	good (1.3–260 ppm)	yellow → orange >conc. →	sol[f]	sol[f]	sl sol[f]	v sl sol[f]	insol[f]	good
Titanium dioxide[g] (inorganic pigment)	exc	exc	exc	exc	exc	fair (50–5,000 ppm)	white	insol[h]	insol	insol	insol	insol	good
Turmeric extract and oleoresin	poor	mod	poor		good	good (Ext., 0.2–60 ppm; Oleo., 2–640 ppm)		sl sol		sol	sol	sol	fair

[a]Prop Gly = propylene glycol.
[b]Not to exceed 15 mg/lb, or pint, of food.
[c]Moderate dispersibility in water.
[d]Not to exceed 30 mg/lb, or pint, of food.
[e]For coloring beverages.
[f]Solubility data for crocin, component of saffron.
[g]Not to exceed 1% by weight of the food.
[h]Good dispersibility in water.

Appendix A
Definition of Terms Pertaining to Colors

A color additive, as defined by Sec. 201(t) of the Federal Food, Drug, and Cosmetic Act as Amended, is "a dye, pigment, or other substance made by a process of synthesis or similar artifice, or extracted, isolated, or otherwise derived, with or without intermediate or final change of identity, from a vegetable, animal, mineral, or other source," that, "when added or applied to a food, drug, or cosmetic, or to the human body or any part thereof, is capable (alone or through reaction with other substance) of imparting color thereto." The FD&C Act states that black, white, and intermediate grays are considered colors. A substance such as an agricultural chemical or a soil or plant nutrient, which is used solely to affect the growth or other natural physiological processes of agricultural commodities, is not a color additive, even though it may affect coloring.

A straight food color, frequently referred to as a primary food color, is a color additive listed in Subparts C and D of Title 21 Code of Federal Regulations Part 8, including lakes and such substances as are permitted by the specifications for such color. Subpart C designates for food use color additives that are subject to certification; Subpart D designates those exempt from certification.

The term lake means a straight color extended on a substratum by adsorption, coprecipitation, or chemical combination that does not include any combination of ingredients made by simple mixing processes.

The certification procedure requires that the manufacturer submit a representative sample from each batch of a straight food color to the Food and Drug Administration for chemical analysis. If the batch complies with the FDA specifications in all respects, a lot number is assigned for identification purposes. This ensures that the batch tested is virtually identical with the material that had been used in animal feeding studies during the toxicologic testing program on which approval of the color was based. Full records must be kept to account for all sales against the lot number assigned.

A mixture indicates a color additive made by mixing two or more straight colors, or one or more straight colors and one or more approved diluents.

A diluent is any component of a color additive mixture that is not itself a color additive and that has been intentionally introduced to facilitate the use of the mixture in coloring foods. The diluent may serve another functional purpose in the foods, such as sweetening, flavoring, emulsifying, or stabilizing the product.

The term FD&C color is a carry-over from regulations promulgated under the Federal Food, Drug, and Cosmetic Act of 1938; it refers to synthetic colors that were regarded as "harmless and suitable for use" in foods, drugs, and cosmetics. The nomenclature has been retained to avoid confusion, but the Color Additives Amendment of 1960 provides for separate listing of colors for use in foods, drugs, and cosmetics, respectively.

Appendix B Chronological History of Synthetic Colors in the United States

Year Listed for Food Use	Common Name	FDA Name	Color Index No.	Year Delisted	Currently Permitted in Food
1907	Ponceau 3R	FD&C Red No. 1	16155	1961[a]	no
1907	Amaranth	FD&C Red No. 2	16185	—	yes[m]
1907	Erythrosine	FD&C Red No. 3	45430	—	yes
1907	Orange I	FD&C Orange No. 1	14600	1956[b]	no
1907	Naphthol Yellow S	FD&C Yellow No. 1	10316	1959[c]	no
1907	Light Green SF Yellowish	FD&C Green No. 2	42095	1966[d]	no
1907	Indigotine	FD&C Blue No. 2	73015	—	yes
1916	Tartrazine	FD&C Yellow No. 5	19140	—	yes
1918	Sudan I	—	12055	1918	no
1918	Butter Yellow	—		1918	no
1918	Yellow AB	FD&C Yellow No. 3	11380	1959[e]	no
1918	Yellow OB	FD&C Yellow No. 4	11390	1959[f]	no
1922	Guinea Green B	FD&C Green No. 1	42085	1966[d]	no
1927	Fast Green FCF	FD&C Green No. 3	42053	—	yes
1929	Ponceau SX	FD&C Red No. 4	14700	—	yes[g]
1929	Sunset Yellow FCF	FD&C Yellow No. 6	15985	—	yes
1929	Brilliant Blue FCF	FD&C Blue No. 1	42090	—	yes
1939	Naphthol Yellow S potassium salt	FD&C Yellow No. 2	10316	1959[h]	no
1939	Orange SS	FD&C Orange No. 2	12100	1956[i]	no
1939	Oil Red XO	FD&C Red No. 32	12140	1956[j]	no
1950	Benzyl Violet 4B	FD&C Violet No. 1	42640	—	yes
1959	Citrus Red No. 2	Citrus Red No. 2	12156	—	yes[k]
1966	Orange B	Orange B	—	—	yes[l]
1971	—	FD&C Red No. 40	—	—	yes

[a]Delisted for food use in 1961 but permitted in externally applied drugs and cosmetics under the name Ext. D&C Red No. 15; delisted completely in 1966.
[b]Delisted for food use in 1956 but permitted in externally applied drugs and cosmetics under the name Ext. D&C Orange No. 3; delisted completely in 1968.

[c]Delisted for food use in 1959 but permitted in externally applied drugs and cosmetics under the name Ext. D&C Yellow No. 7.
[d]Deleted for food, drug, and cosmetic use in 1966 because of lack of economic importance.
[e]Delisted for food use in 1959 but permitted in externally applied drugs and cosmetics under the name Ext. D&C Yellow No. 9; delisted completely in 1960.
[f]Delisted for food use in 1959 but permitted in externally applied drugs and cosmetics under the name Ext. D&C Yellow No. 10; delisted completely in 1960.
[g]Food use limited to the coloring of maraschino cherries, at a level not exceeding 150 ppm by weight.
[h]Delisted for food use in 1959 but permitted in externally applied drugs and cosmetics under the name Ext. D&C Yellow No. 8; delisted completely in 1962.
[i]Delisted for food, drug, and cosmetic use in 1956 but subsequently relisted as Ext. D&C Orange No. 4; delisted completely in 1963.
[j]Delisted for food use in 1956 but permitted in externally applied drugs and cosmetics under the name Ext. D&C Red No. 14; delisted completely in 1963.
[k]Food use limited to the coloring of skins of oranges not intended or used for further processing, at a level not exceeding 2 ppm by weight.
[l]Food use limited to the coloring of casings and surfaces of frankfurters and sausages, at a level not exceeding 150 ppm by weight.
[m]The status of FD&C Red No. 2 is under review.

References

Certified Color Industry Committee. 1968. Guidelines for good manufacturing practice: Use of certified FD&C colors in food. Food Technol. 8:946.

Davis, K. J., O. G. Fitzhugh, and A. A. Nelson. 1964. Chronic rat and dog toxicity studies on tartrazine. Toxicol. Appl. Pharmacol. 6:621.

Davis, K. J., A. A. Nelson, R. E. Zwickey, W. H. Hansen, and O. G. Fitzhugh. 1966. Chronic toxicity for Ponceau SX in rats, mice, and dogs. Toxicol. Appl. Pharmacol. 8:306.

Hansen, W. H. 1962. Chronic studies on FD&C colors. Bureau-By-Lines 3:23.

Hansen, W. H., O. G. Fitzhugh, A. A. Nelson, and K. J. Davis. 1966. Chronic toxicity of two food colors. Brilliant Blue FCF (FD&C Blue No. 1) and Indigotine (FD&C Blue No. 2) in rats and dogs. Toxicol. Appl. Pharmacol. 8:29.

Hesse, B. C. 1912. Coal-tar colors used in food products. Bureau of Chemistry, U.S. Dep. Agr. Bull. 147.

Joint FAO/WHO Expert Committee on Food Additives. 1963. Specifications for identity and purity of food additives: Food colors, Vol. II.

Joint FAO/WHO Expert Committee on Food Additives. 1965. Specifications for the identity and purity of food additives and their toxicological evaluation: Food colours and some antimicrobials and antioxidants. FAO Nutr. Meet. Rep. Ser. No. 38; WHO Tech. Rep. No. 309.

Joint FAO/WHO Expert Committee on Food Additives. 1966. Specifications for identity and purity and toxicological evaluation of some food colours. FAO Nutr. Meet. Rep. Ser. No. 38B; WHO/Food Add/66.25.

Joint FAO/WHO Expert Committee on Food Additives. 1967. Specifications for the identity and purity of food additives and their toxicological evaluation: Some emulsifiers and stabilizers and certain other substances. FAO Nutr. Meet. Rep. Ser. No. 43; WHO Tech. Rep. Ser. No. 373.

Joint FAO/WHO Expert Committee on Food Additives. 1970a. Specifications for the identity and purity of food additives and their toxicological evaluation: Some food colours, emulsifiers, stabilizers, anticaking agents, and certain other substances. FAO Nutr. Meet. Rep. Ser. No. 46; WHO Tech. Rep. Ser. No. 445.

Joint FAO/WHO Expert Committee on Food Additives. 1970b. Toxicological evaluation of some food colours, emulsifiers, stabilizers, anticaking agents and certain other substances. FAO Nutr. Meet. Rep. Ser. No. 46A; WHO/Food Add/70.36.

Joint FAO/WHO Expert Committee on Food Additives. 1970c. Specifications for the identity and purity of some food colours, emulsifiers, stabilizers, anticaking agents and certain other substances. FAO Nutr. Meet. Rep. Ser. No. 46B; WHO/Food Add/70.37.

ISBN 0-309-019